G. Hermann Jonas TAPSOBA

Virulence of Candida tropicalis : Experimental study

G. Hermann Jonas TAPSOBA

Virulence of Candida tropicalis : Experimental study

Study of the virulence of a telluric strain of Candida tropicalis using the Galleria mellonella model

ScienciaScripts

Imprint

Cover image: www.ingimage.com

This book is a translation from the original published under ISBN 978-620-6-71345-6.

Publisher:
Sciencia Scripts
is a trademark of
Dodo Books Indian Ocean Ltd. and OmniScriptum S.R.L publishing group

120 High Road, East Finchley, London, N2 9ED, United Kingdom
Str. Armeneasca 28/1, office 1, Chisinau MD-2012, Republic of Moldova, Europe
Managing Directors: Ieva Konstantinova, Victoria Ursu
info@omniscriptum.com

Printed at: see last page
ISBN: 978-620-8-62463-7

DEDICATION

I dedicate this work to :

✓ To God, Heavenly Father!

To you, to whom I owe this breath of life, I give thanks for the health and strength you give me night and day to help those in need of care.

✓ To my late father, Captain Joseph M. P. TAPSOBA, In Memoriam

Dad, I'll always carry you in my pious thoughts. Be proud of your son.

✓ To my beloved mother, Marcelline KOAMA

You spare no effort to support your children morally, physically and spiritually. May the Heavenly Father grant you long life so that we may always benefit from your advice.

✓ To my little family: my wife Aissata OUEDRAOGO, my children Syntiche Aurélia and Naelle Assumpta*. You are the reason for my socio-professional struggle; I love you.*

ACKNOWLEDGEMENTS

This work is the fruit a constellation of support in my person; My sincere thanks

√ ***To my sisters and brother*** *Armelle, Nadia, Floriane, Félicia, Alain*

√ ***To the staff of the Parasitology-Mycology Department at the CHU-YO**, who were always courteous and available to help me with my training.*

√ ***To Dr Marcel P. SAWADOGO,*** *Pharmacien Parasitologiste, who co-supervised this work, for his simplicity and availability. Thank you for your guidance.*

√ ***To Dr Adama ZIDA (MCA)**, Pharmacist Parasitologist, Head of the Parasitology-Mycology Department of the CHU/YO and Director of our dissertation, for his availability and his known rigour in the work. Thank you for your guidance.*

√ ***Dr Clément KAFANDO,*** *pharmacist, for his dynamism and technical support for this work.*

√ ***To my friends, colleagues and other collaborators**.*

TO OUR MASTERS AND JUDGES

To our teacher and supervisor, ***Dr Adama ZIDA (MCA)***

You are

- ***Pharmacist, Former intern at the Hospitals of Ouagadougou;***
- ***Associate Lecturer in Parasitology-Mycology at the URF/SDS of Professor Joseph Ki-ZERBO University;***
- ***Head of the Parasitology-Mycology Department at CHU-YO ;***
- ***Technical contact for neglected tropical diseases.***

Dear Master,

The little story of complicity between you and me goes back ten years now, when I knocked on the door of your office and you welcomed me with open arms, when I was a trainee intern at the end of my course in General Medicine, passionate about Parasitology-Mycology and you were a young Head of Department. You didn't hesitate for a to supervise me day and night so that my doctoral thesis could be accepted with great honour. In the same vein, you were our mentor for this work. We admire your passion and rigour for science. What's more, you're a confidant to us, always ready to listen and offer helpful advice. We would like to take this opportunity to congratulate you on your success in the CAMES agrégation competition in 2020.

May the Lord grant you and your family health, peace and happiness.

To our teacher and dissertation co-director, ***Dr Patindoilba Marcel SAWADOGO***

You are

- ***Former intern at the Ouagadougou Hospitals,***
- ***Assistant in Parasitology-Mycology at the Health Sciences Training and Research Unit (UFR/SDS) of Joseph KI-ZERBO University,***
- ***WHO expert in malaria microscopy.***

Dear Master,

We didn't hesitate for a moment to approach you to discuss the subject of our current memoir. It was a real pleasure for us to work with you on this project in a friendly atmosphere, which was rich in exchanges, advice and opinions. Your humility, your simplicity, your high sense of caring for your neighbours make you a very admirable character.

May the Lord grant you and your family health and happiness.

TABLE PF CONTENTS

INTRODUCTION

Over the last three (03) decades, there has been a significant increase in the incidence of fungal infections in humans [1]. These infections may be superficial, affecting the skin, hair, nails and mucous membranes, or systemic, involving several organs of the body [2]. A number of factors appear to be involved in this increased frequency of fungal infections, but it is generally accepted that the increase and inappropriate use of certain medical practices such as immunosuppressive therapies, invasive surgical procedures and the use of broad-spectrum antibiotics are important [3, 4, 5]. Among human pathogenic fungi, those of the Candida genus are the most commonly encountered. The Candida genus contains more than 150 heterogeneous species [6]. Candida albicans has been described as the species most frequently associated with fungal infections in humans. However, Candida species other than C. albicans have become an important cause of deep and systemic candidiasis in nosocomial infections, including C. tropicalis. Worldwide, C. tropicalis accounts for 3 to 66% of Candida blood isolates [7, 8, 9] and in many hospitals is the second most common Candida species. This variability in prevalence depends on geographical location: USA 12 to 25%, Europe 4.5 to 9%, Brazil 20 to 24% [10,11,12].In India, 67 to 90% of cases of candidemia were due to species other than C. albicans. and C. tropicalis was the most dominant [13, 14]. In tropical regions, C. tropicalis remains a major cause yeast infections. In Abidjan (Ivory Coast), C. tropicalis and C. albicans are the main causes of vulvovaginitis [10]. In Burkina Faso in 2017, Sangaré et al[15] isolated Candida tropicalis in Bobo Dioulasso as the second most common non-albicans Candida species associated with candidiasis vulvovaginitis in pregnant women[11]. Several animal methods can be used to study the pathogenicity of Candida: invertebrate systems and vertebrate systems. Invertebrate systems such as Galleria mellonella,

Caenorhabditis elegans, Bombyx mori and Drosophila melanogaster offer a number advantages over vertebrate systems, mainly because they allow the study of strains without the ethical considerations associated with studies on invertebrates [16].

We chose the insect model G. mellonella because of its ease of handling and culture. This model gave results comparable to those obtained using mammalian models [16].

Recent evidence suggests that the immune system of G. mellonella is good compared to other invertebrates and is similar to that of vertebrates, as it has both a cellular and humoral immune response to infection [16].

In addition, Galleria mellonella larvae evolve in an aerobic environment at temperatures between 15° C and 37°C, making it possible to study the pathogenicity of yeasts at human body temperature, whereas other invertebrates cannot generally be maintained at temperatures above 30°C [17].

In Burkina , studies on the virulence of Candida using the invertebrate model are at a preliminary stage. The aim of our study is to assess the in vivo virulence of a telluric strain of C. tropicalis using the G. mellonella model.

PART I

GENERAL INFORMATION

I. Candida tropicalis

I.1. Characteristics microbiological

I.1.1 Taxonomy / Classification

The classification of fungi has evolved considerably. It dates back to 2000 [18]:

Kingdom: Mushrooms

Phylum : Ascomycota

Class: Hemiascomycetes

Order: Saccharomycetales

Family: Candidaceae

Genus: Candida (Berkhout, 1923)

The Candida genus comprises around 200 species, the most common of which are found in human pathology: C. albicans, **C. tropicalis**, C. glabrata, C. krusei, C. guilliermondii, C. parapsilosis, C. kefyr and C. dubliniensis [19, 20].

I.1.2. Morphological and cultural characteristics

C. tropicalis is a non-capsulated, non-pigmented, aerobic yeast. On Tween Agar 80 medium, after 72 hours incubation at 25°C, it produces oval blastospores (4-8mm) with the occasional formation of pseudomycelia and even true mycelia [21]. On CHROM-agar, C. tropicalis colonies appear dark blue.

Table I shows the distinctive morphological characteristics of C. tropicalis and other Candida species.

Table I: Distinctive morphological features between C. tropicalis and other Candida species [22].

Species	Tube of germination	Mycelium/ pseudomycelium	Yeast size (µm)	Colour colour on CHROM-agar
C. tropicalis	-	± /+	4-8 x 5-11	Dark blue
C. albicans	-	+ /+	4-6 x 6-10	Bluish green
C. parapsilosis	-	- /+	2,5-4 x 2,5-9	white
C. glabrata	-	- / -	1-4	White, purple

I.2. Pathogenicity and virulence factors

Virulence factors are those that interact directly with host cells, causing damage. Pathogenicity is mediated by a number of virulence factors [23]. Several virulence factors have been proposed:

- **Adhesion and biofilm formation**: The fungal cell surface is the site of interaction with host tissues or medical devices, leading to yeast adhesion. This adhesion is mainly due to membrane proteins of the Agglutinin-like sequences (ALS) family. These proteins bind to specific receptors on the surface of epithelial cells (for example E-Cadherin and N-Cadherin for ALS 3 [24]. By binding to cadherins, these adhesion molecules can also induce endocytosis of the yeast or filament by the epithelial cells. Inactivation of the expression of these proteins by C. tropicalis is responsible for a reduction in adhesion and virulence in murine models systemic and localised infection [25, 26]. Adhesion to the epithelium immediately triggers filamentation [27] and the secretion of proteases (Secreted Aspartyl proteases (SAP5 and SAP9)). These promote the penetration of mycelial filaments into tissues.

- **Pseudomycelium formation**: in general, the transition from the yeast form to the more or less filamentous form is associated with virulence. This filamentation allows C. tropicalis to invade tissues and reach the general circulation. From this point onwards, the reversion of the filamentous form to a yeast form will enable it to spread at a distance and establish new outbreaks.

- **Interference with phagocytosis**: C. tropicalis is capable of producing acid peptides that can inhibit phagocytosis binding and oxidative metabolism. In addition, the yeast can induce apoptosis in macrophages and neutrophils, thus escaping the cells of the immune system [28,29].
- **Interference with complement**: Fungal adhesins, related to the CR3 receptor on lymphocytes, can be refined for certain plasma matrix components such as fibronectin, but also the C3bi fraction of complement, thus disrupting yeast phagocytosis.
- **Enzymes:** the secretion of hydrolytic enzymes during infection promotes virulence by degrading the mucosal surfaces of the host and its immune defences. These enzymes are aspartyl proteinases (Saps), phospholipases and lipases [30].

II. Galleria mellonella

II.1. Definition

Galleria mellonella is a ubiquitous insect of the Lepidoptera family. It is harmful to colonies European honeybees, Apis mellifera (L.), and was used as a non-mammalian animal model in biomedical research in the late 1990s and early 2000s [31].

II.2. Classification / Taxonomy

Galleria mellonella belongs to :

Class: insect

Order: butterfly

Family: pyralidae, lepidoptera

Genre: Galleria

Species : Galleria mellonella

G. mellonella larvae can cause significant damage to the wax combs making up the hive by feeding on them [32].

II.3. Cycle

It takes place in 04 stages: egg, larva, pupa and imago or adult butterfly (Figure 1). Under ideal conditions, the cycle lasts only 06 weeks, i.e. 4 to 6 generations per year. This duration is highly dependent on environmental factors and the availability of food [32, 33, 34, 35].

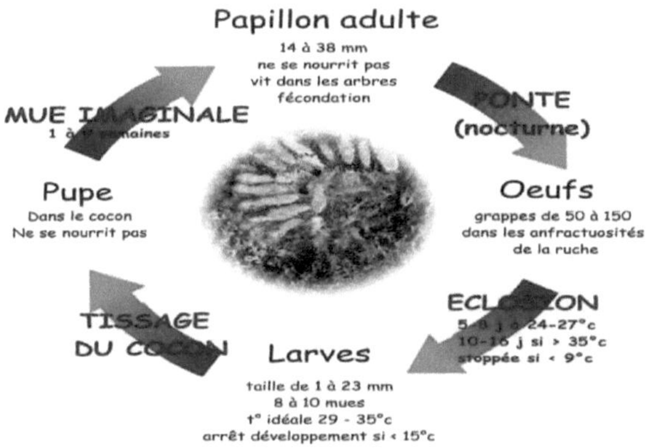

Figure 1: Development cycle of G. mellonella [32].

II.4. Morphology

Figure 2 shows the morphology of G. mellonella at different larval stages:

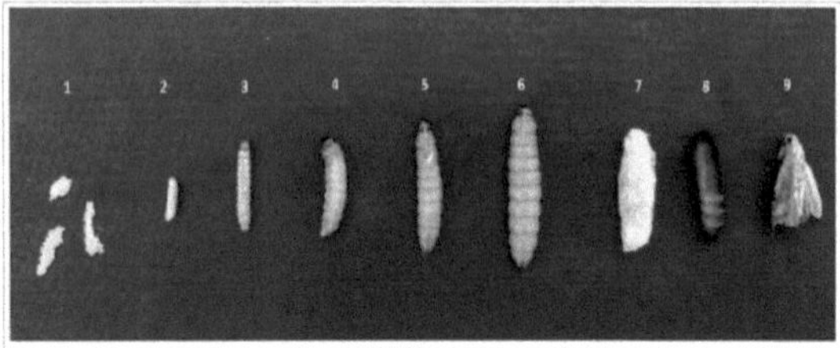

Figure 2: different stages of development of G. mellonella [34]. (1): egg approximately: 10d. (2), 20D (3), 25D. (4), 35J. (5), 40 days (6), prepupa (7), pupa (8) and adult (9). D = days

II.5. Immune system of G. mellonella

Several studies have revealed similarities between the innate immune system of G. mellonella and that of mammals [36]. In insects, the first barrier to infection is the outer envelope or cuticle. As soon as an infectious agent (bacterium, fungus or virus) crosses this wall, it encounters two types of immune response the cellular response and the humoral response [37, 16]. In the haemocoel, the equivalent of mammalian blood, the microorganism is recognised by cells called haemocytes, of which there are 06 types in G. mellonella: plasma cells, granulocytes, spherulocytes, coagulocytes, adipohaemocytes and oenocytordes. Three mechanisms involved in cellular immunity: nodulation, encapsulation and phagocytosis [36]. The humoral response is triggered simultaneously. As in mammals, TLRs (Toll-like receptors), receptors innate immunity expressed by cells in the haemolymph, recognise the foreign element; their activation initiates the production of antimicrobial peptides (AMPs), small molecules capable of acting on a wide range of microorganisms [38, 39].

The most visible mechanism of larval immunity is the production of melanin. Unlike humans, melanin in the larva is the result of a defence process against pathogens. When an infection occurs, a tyrosinase is converted into its active form, phenoloxidase, which triggers a cascade of responses culminating in the production of melanin. This cascade is part of the larval coagulation system that is common to most insects and corresponds, in mammals, to the complement cascade that enables coagulation [36].

II.6. Medical use of G. mellonella

Very recently, G. mellonella larvae have been used to test the pathogenicity of Gram-negative and Gram-positive bacteria (BGN and BGP) [40]. Until now, these tests have been carried out using mammalian models, but the ease with which moths can be used has attracted researchers to this type of analysis. However, this example illustrates the potential use of moths as a model for rapid screening and identification of potentially pathogenic bacteria within a complex population. The characterisation of the genes responsible for the production of virulence factors another field in which larvae can be useful. Researchers at the University of Milan [41] used G. mellonella to study the role of glucose import systems in the virulence of Pseudomonas aeruginosa. The melanisation has a major advantage for the experimenter: it allows him to follow, with the naked eye, the progress of the infection as a function of the change in colour of the caterpillar. The larvae are also capable of living at temperatures between 15°C and 37°C, enabling incubation at the growth temperature of the microorganism under study and/or at human body temperature.The advantages of using G. mellonella, its applications in medicine and the different parameters analysed are summarised **in Figure 3**.

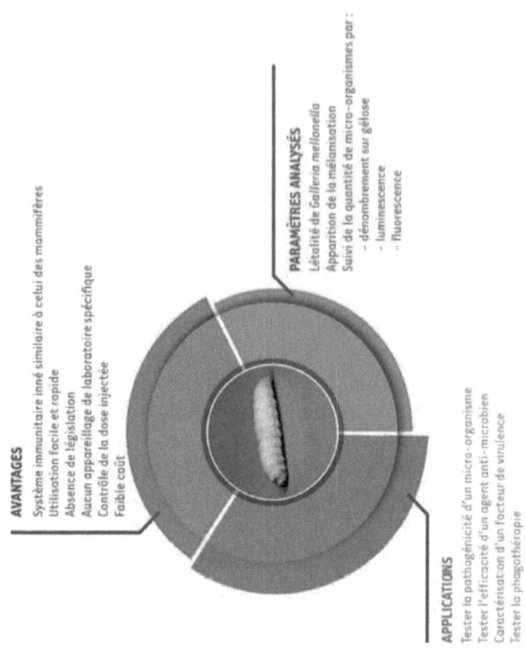

Figure 3: advantages, applications and parameters analysed for G. mellonella [42].

PART II

OUR STUDY

I. Methodology

I.1. Objectives

I.1.1. General objective

To assess the in vivo virulence of the amphotericin B-resistant Candida tropicalis telluric strain using the Galleria mellonella model.

I.1.2. Specific objectives

► Determine the rate of melanisation of G. mellonella larvae after injection of the telluric strain of C. tropicalis ;

► To determine the mortality rate of G. mellonella larvae after injection of the telluric strain of C. tropicalis;

► Estimate the in vivo virulence of the C. tropicalis strain based on the number CFUs.

I.2. Type and period of study

This was a prospective study with the aim of evaluating the in vivo virulence of the telluric strain of C. tropicalis resistant to amphotericin B using the Galleria mellonella model.

The study ran from July to September 2020.

I.3. Materials and methods

I.3.1. Materials

Our study material was :

- **G. mellonella larvae**. They were cultured in the parasitology laboratory of UFR/SDS;

- **The telluric strain of C. tropicalis**: this was selected at random from several strains of C. tropicalis isolated from the soil in Ouagadougou.

I.3.2. Reagents and media for culture

The reagents and culture media were composed as follows:

- Phosphate buffered saline (PBS= Phosphate Buffered Saline),

- Yeast extract Peptone Dextrose (YPD) medium,

- Sabouraud medium+ Chloramphenicol (SC).

I.3.3. Method

I.3.3.1. Culture conditions for G. mellonella larvae

We used forty larvae derived from the reproduction of G. mellonella butterflies from Institut de Recherche en Santé (IRS) 2 at the University of Nantes (France). The conditions under which these larvae were reared were rigorously observed. Healthy larvae weighing between 280 and 320 mg were included in our study. Diseased larvae (i.e. larvae with dark spots or apparent melanisation) and those weighing less than the required amount were not included.

I.3.3.2. Method for identifying strains of C. tropicalis

Soil samples were taken from various sites arrondissement N°3 of the city of Ouagadougou **(Figure 4)**. The GPS coordinates of the site from which our strain were "12°20'04.9"N 1°34'13.2"W". The various yeast strains, including C. tropicalis were isolated using the following protocol **(Figure 5)**:

Figure 4: Geographical location of soil sampling sites in the city of Ouagadougou

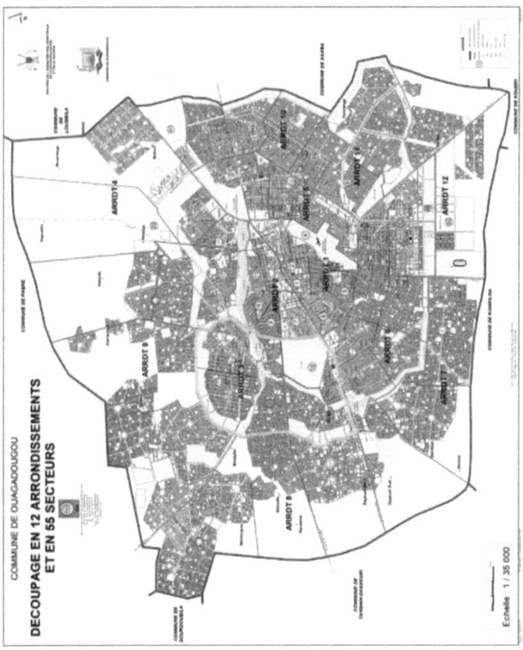

Source: Image Dr Sawadogo P. Marcel

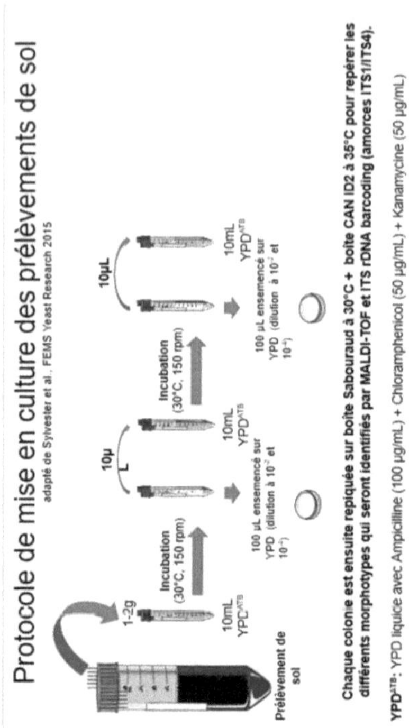

Figure 5: Culture protocol for soil samples.

The different strains were seeded in YPD medium (1% yeast extract, 2% peptone and 2% dextrose) + Chloramphenicol and incubated at 30°C in a shaking incubator followed subculturing in Sabouraud medium + Chloramphenicol. Each colony obtained by culture was then subcultured onto Sabouraud medium at 30°C and then onto CAN ID2 medium at 35°C **(figure 6)** to identify the different Candida morphotypes. The Candida morphotypes were subjected to Matrix Assisted Laser Desorption Ionization Time-Of-Flight (MALDI-TOF) for strain identification **(figure 7)**. Only one strain (S1R1Y1) of C. tropicalis was included in our study.

Figure 6: Candida colonies subcultured on CAN ID2 medium

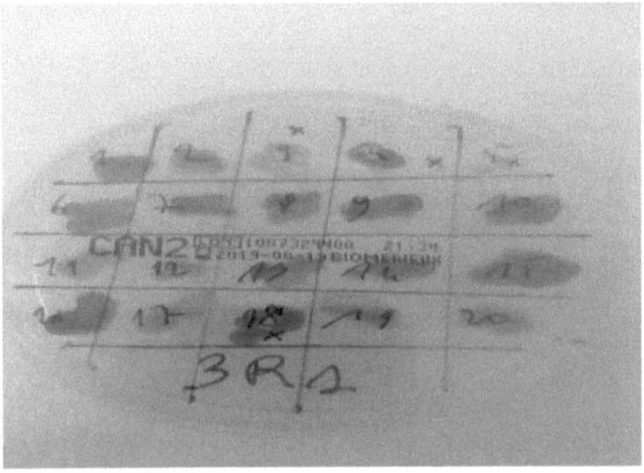

Source: photos by Dr M. Sawadogo, 2019

Figure 7: reporting of Candida strains identified by MALDI-TOF

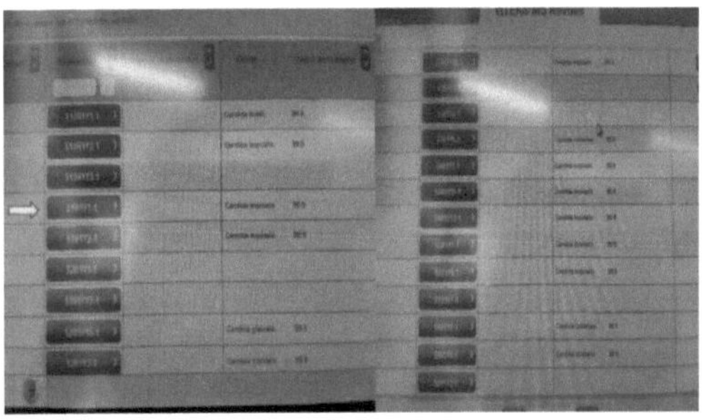

Source : Photos Dr M. SAWADOGO, Nantes 2019

I.3.2.3. Preparation of cell suspensions

The different strains of C. tropicalis grown after 24 hours were collected and used to prepare the cell suspensions. Each strain was suspended in 1000µl of PBS

I.3.2.4. Inoculation of G. mellonella with Candida tropicalis strain

Larvae were selected for injection on the basis of their weight (280 to 320 mg) and placed in sterile Petri dishes. Batches of 10 healthy larvae were made up and disinfected with 70% alcohol at the injection sites, i.e. the left or right last extension. The larvae were inoculated with 20µl/larva of cell suspension at increasing concentrations of 5.10^5, 1.10^6 and 5.10^6 colony-forming units (CFU) of the strain using a fine-needle insulin syringe **(Figure 8)**. The inoculum was prepared with a PBS solution. A control group received 20µL/larve of PBS and the other 03 at the above concentrations. After injection, the larvae were incubated at 37°C without food. Melanisation (generally occurring between 30 and 60 min after inoculation) was observed as well as mortality every 24 h. The day natural post-infection death was recorded for each larva. Larval death was assessed by the lack of movement of the larvae in response to stimulation associated with discolouration of the cuticle. Dead larvae were removed from the Petri dish containing the remaining viable larvae. Average mortality was calculated per day for each type of strain (non-resistant and resistant strains) for each inoculum. The data obtained were used to determine survival curves. The survival experiments ended 12 days after infection. A strain is considered to be virulent if it causes 100% mortality in the larvae in a maximum of 96 hours and at a given value of CFU inoculated.

Figure 8: Method injecting a G. mellonella larva into the last proleg.

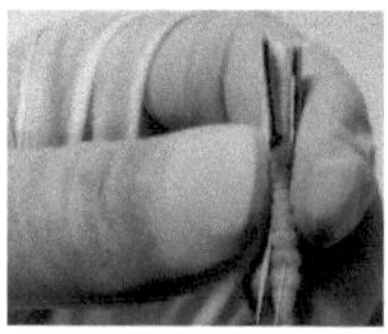

Source: Photo of UJKZ Parasitology-Mycology Laboratory

I.4. Ethical considerations

Galleria mellonella caterpillars are not subject to the regulations applied to vertebrates.

I.5. Processing and analysis of data

For the virulence study, mortality curves were plotted and examined using the Kaplan-Meier method on XLSTAT 2020.1.2 (Addinsoft) for EXCEL and differences were determined using the log-rank test. The difference was considered significant at $p<0.05$.

II. Results

II.1. Response of C. tropicalis S1R1Y1 to antifungal agents

The S1R1Y1 strain of C. tropicalis was resistant to amphotericin B but sensitive to Fluconazole. Table II shows the response of the C. tropicalis strain to antifungal agents.

Table II: Response of C. tropicalis strain S1R1Y1 to antifungal agents

C. tropicalis S1R1Y1 strain

Antifungal	that	S	I	R
AMB		--	--	R
FLC		S	--	--
MIC		S	--	--
KCA		S	--	--
NS		S	--	--
OCOL		S	--	--
ECN		S	--	--
MIC		S	--	--
S: sensitive ;	I: intermediate ;	R: resistant ;	AMB: amphotericin B ;	FLC: fluconazole ;

MIC: miconazole; KCA: ketoconazole; NS: nystatin; CLO: clotrimazole; ECN: Econazole.

Figure 9 shows an illustration of the antifungaligram for the C. tropicalis strain

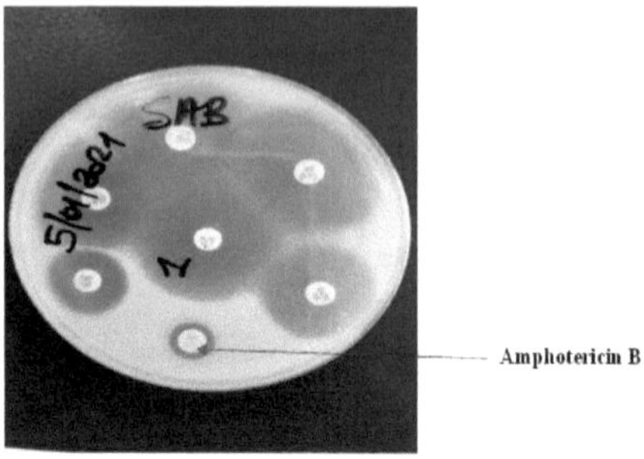

Figure 9: Antifungiogram using the C. tropicalis strain

II.2. Study of the virulence of the telluric strain of C. tropicalis according to the quantity CFU of C. tropicalis

In vivo virulence tests should be able to distinguish the most virulent concentrations from those with little or no virulence.

II.2.1. Melanisation thirty (30) minutes after inoculation

One hundred per cent of the larvae in batches infected at high concentrations melanised 30 minutes after inoculation. Table III shows the melanisation rate of larvae infected with the different strains of C. tropicalis.

Table III: Melanisation rate of G. mellonella 30 minutes after inoculation of suspensions of the strain at different concentrations.

Groupe de G. *mellonella* par niveau de concentration	Taux de larves mélanisées (%)
G1 à 5.10^5	80
G2 à 1.10^6	100
G3 à 5.10^6	100
PBS (contrôle)	0

G = Groupe

II.2.2. Melanisation one (1) hour after inoculation

We found that, with the exception of the group inoculated with PBS, all G. mellonella larvae melanised one (1) hour after inoculation.

II.2.3. Melanisation two (02) and twenty-four (24) hours after inoculation

The results were similar to those obtained after the first hour. The batch of larvae that had received the PBS remained intact **(figure 10).**

Figure 10: Physical changes in G. mellonella larvae 1H after inoculation with different strains of C. tropicalis

Source: Photo of UJKZ Parasitology-Mycology Laboratory

A, the control and the batch injected with PBS showed no melanisation.

B, lot with a fungal load of 5.10^6 CFU/plant; abundant melanisation is noted

II.2.4. 12-day survival of G. mellonella larvae inoculation of the suspension of strain S1R1Y1

Survival was compared with the batch of larvae given only injection with PBS.

Figure 11 summarises the survival rate of the larvae as a function of time.

Figure 11: Survival of groups of G. mellonella larvae after infection with different C. tropicalis inocula.

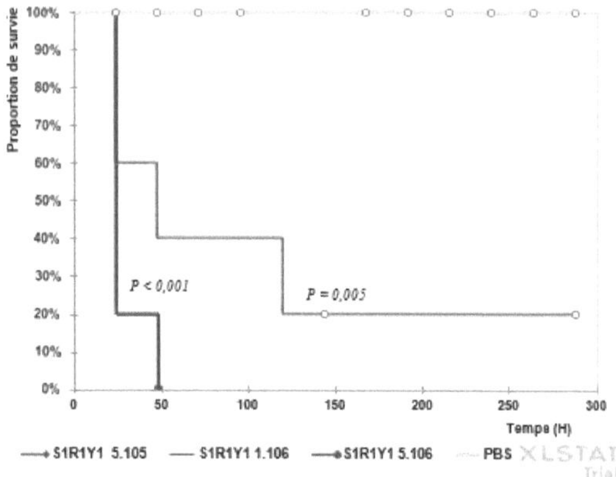

No mortality was observed in the batch injected with PBS.

For inoculum concentration of 5.10^5 CFU, we observed an average survival time of 67.20 ± 14.9 hours ($p<0.05$) and a survival rate of 10% of the batch of larvae at the end of 12 days. The mortality rate observed in the batch that received the 1.10^6 CFU inoculum was similar to that of the batch that received the 5.10^6 CFU inoculum, the average survival time for these two batches being 28.8± 3.2H ($p<0.05$). For both batches, 100% mortality was observed after 48 hours.

III. Discussion

III.1. Limitations of our study

The limitations and difficulties encountered in the course of this study were :

Insufficient numbers of G. mellonella larvae, which had the following disadvantage:

- The restriction on the number of larvae per batch set up during the experiment. We could have set up batches with a large number of larvae for better sampling.
- Non-repeatability of the experiment: ideally, the experiment should be repeated at least twice under the same conditions.
- The limitation in the selection of strains; in fact only the S1R1Y1 strain of C. tropicalis was used during the experiment. It would have been more interesting to use several strains from this species in order to compare their virulence.
- The limitation in the diversification of yeast suspension concentrations during the experiment; intermediate concentrations could have been used to better assess the virulence of the concentration-dependent strain.
- The study of sensitivity to antifungals carried out in vitro; the larval sensitivity profile to antifungals could have been determined by simultaneously injecting yeast suspensions at the concentrations used and antifungals into the larvae and observing the survival time of the larvae.

III.2. Choice of the G. mellonella model for our study

The best-known traditional model, which has been used for the longest time, is the mouse model. Various mouse models have therefore been developed for experimental studies with microorganisms.However, ethical, economic and logistical considerations limit the use of mammals in infection experiments, particularly when the question posed requires a large number of guinea pigs. Alternative approaches have been developed, in particular using Caernorhabditis elegans, Drosophila melanogaster and G. mellonella larvae as hosts [42]; The use of G. mellonella larvae is interesting in that : Breeding G. mellonella is not technically demanding, either in terms of financial plan ;

The relatively large size of the larvae makes them easy to handle; Larvae can be maintained at temperatures close to those of mammals, which is not the case for other invertebrates. It has been shown that temperature affects the expression of Candida virulence traits [28] ;

The larval immune system has functional and structural similarities with the innate immune system of mammals [16, 36, 37, 38, 39].

In comparison with the murine infection model, frequently described on the basis of experiments carried out on mice, Brennan et al. in 2002 [51] in Montreal reported that there was a strong correlation in the virulence of C. albicans strains between the murine models and G. mellonella. Different concentrations of thirteen strains of C. albicans inoculated into groups of these two hosts showed this strong correlation.

III.3. Response of C. tropicalis S1R1Y1 to antifungal agents

The telluric strain of C. tropicalis was resistant to AMB and sensitive to FLC; AMB, a polyene, binds covalently to ergosterol to form pores in the plasma membrane through which electrolytes pass uncontrollably and cause the death of the fungus [43]. Fitzpatrick et al in 2006, Mesa-Arango et al in 2014, Young et al in 2003, [23, 44, 45] in their studies on the mechanisms of resistance of C. tropicalis strains to AMB showed that this resistance was associated with various changes within the yeast; they noted a significant decrease in the amount ergosterol in the membrane of the yeast. due to a mutation in the ERG genes. In addition, transmission electron microscopy (TEM) showed that the cell wall of these strains was abnormally thickened. We assume that the resistance conferred on strain S1R1Y1 to AMB is acquired resistance, given that most strains of C. tropicalis are naturally sensitive to AMB; this could be explained by the irrational and use of AMB.

III.4. Melanisation of larvae after inoculation

No melanisation was observed within 24 hours of PBS injection. However, the opposite observation was made with batches of larvae inoculated at different concentrations with the C. tropicalis strain. Melanisation was observed in 100% of these larvae one hour after injection of the inoculum. Studies [16, 47] suggest that the melanisation observed is due to the formation and accumulation of melanin in the larval haemolymph. This melanisation reflects the activation of the immune system in response infection. The melanisation response can be described as the synthesis and deposition of melanin to encapsulate the yeast at the wound site, followed by haemolymphocoagulation similar to the mammalian complement system and opsonisation [17]. In addition, observation of larvae injected with PBS over 24 hours shows that the breach caused by the syringe needle does not lead to melanisation. Shokal et al in 2017 [48] accept that although melanisation is an essential component of the G. mellonella immune system, it must tightly regulate specific protease inhibitors to allow precise deposition of melanin to circumscribe infection and reduce yeast toxicity; We believe that at relatively high inocula, this regulatory mechanism becomes overwhelmed and melanisation tends to extend throughout the larval haemolymph.

III.5. Estimation of the virulence of the C. tropicalis strain according to the quantity of CFU

The survival time and mortality rate of the larvae inoculum-dependent concentration were used as virulence criteria for the C. tropicalis strain and the minimum and maximum inocula were 5.10^5 and 5.10^6 CFU/larva. These concentrations are close to those reported in the literature for experimental studies using the G. mellonella. The PBS solution was chosen as a control for the injection of larvae in accordance with the insect infection model experiments described in the literature. The mean survival for the batch of larvae inoculated with 5.10^5 CFU was longer than that observed with the 10^6 and 5.10^6 CFU inocula (67.20 ± 14.9H and 28.8± 3.2H respectively) and 10% of the larvae survived beyond 96 hours; C. tropicalis strain S1R1Y1 is therefore not virulent at a concentration of 5.10^5 CFU for G. mellonella larvae. Mesa-Arango et al. in Colombia in 2013 [49] in their experiment with clinical strains of C. tropicalis found a longer mean larval survival time, 211.2 ± 28.8 H and 151.2 ± 28.8 H for concentrations of 5.10^5 and 10^6 respectively. The low virulence observed at low inoculum concentrations could be explained by the efficient phagocytosis of the yeast by G. mellonella haemocytes. Mesa-Arango et al [49] suggested, after comparing haemocyte density before and after injection of PBS and C. tropicalis suspensions, that haemocyte numbers fell significantly at high concentrations within 2 to 7 hours.Our results are similar to those obtained by Kafando in 2020 [50] in Ouagadougou who found larval survival of more than 200 hours at a concentration of 5.10^5 but with several clinical strains of C. albicans. The relationship between larval mortality rate, survival time and inoculum concentration was statistically significant ($p<0.05$). Overall, a minimum concentration of 1.10^6 CFU was sufficient to cause 100% lethality. The telluric strain S1R1Y1 is therefore virulent for larvae at a minimum concentration of 1.10^6 CFU.

CONCLUSION

Candida tropicalis has accounted for a significant proportion of Candida species implicated in deep-rooted candidiasis and candidemia in tropical regions in recent decades. This explains its association with high mortality rates in patients, especially those hospitalised in ICUs.Amphotericin B, which has long been the antifungal agent of choice for the treatment of deep-rooted candidiasis, is showing resistance to the Candida tropicalis strain in our experience. G. mellonella infection model has recently proved to be a suitable model for analysing fungal virulence factors as an alternative to experiments on mammals, for which bioethics are a barrier. In our context, this model invertebrate infection, like that of G. mellonella, could be used as the 'Gold Standard' for comparative studies of the virulence of different species within the same genus of fungi or bacteria, or even of the virulence of strains of the same species.

SUGGESTIONS

At the end of our study, we would like to make a few suggestions to the authorities, medical staff, pharmacists and patients.

► The authorities (UFR/SDS hospital management and the Ministry of)

• To strengthen the technical facilities of the Parasitology-Mycology Department of the CHU-YO for more effective diagnosis of candidiasis;

• Provide ongoing training for medical staff in the management of candidiasis;

• To better equip the UFR/SDS Parasitology-Mycology laboratory for improved entomological culture.

► Medical staff

• Avoid prescribing antifungal agents before any antifungals

► Pharmacists

• Dispense antifungals only on presentation of a doctor's prescription

► To patients

• Avoid self-medication with antifungals;

• Strictly respect the advice, therapeutic indications and instructions given by healthcare providers.

REFERENCES

1. **Lass-Florl C.** The changing face of epidemiology of invasive fungal disease in Europe. Mycoses. 2009, 52(3), 197-205.

2. **Rüping M.J.G.T., Vehreschild J.J., & Cornely O.A.** Patients at High Risk of Invasive Fungal Infections. Drugs. 2008, 68(14), 1941-1962.

3. **Samaranayake L., Fidel P., Naglik S., Teanpaisan R., Coogan M., Wanzala P.** Fungal infections associated with HIV infection. Oral Diseases. 2002, 8(s2), 151-160.

4. **Hagerty J.A., Ortiz, J., Reich D., & Manzarbeitia C.** Fungal Infections in Solid Organ Transplant Patients. Surgical Infections. 2003, 4(3), 263-271.

5. **Kojic E.M., Darouiche R.O.** Candida Infections of Medical Devices. Clinical Microbiology Reviews. 2004, 17(2), 255-267.

6. **Calderone R., Latgé J.-P**. Host-microbe interactions: fungi invasive human fungal opportunistic infections. Current Opinion in Microbiology. 2002, 5(4), 355-358.

7. **Bassetti, M., Righi E., Costa A., Fasce R., Molinari M.P., Rosso R., Viscoli C.** Epidemiological trends in nosocomial candidemia in intensive care. BMC Infectious Disease. 2006, 6(1).

8. **Diekema, D., Pfaller M., Jones R.** (2002). Age-related trends in pathogen frequency and antimicrobial susceptibility of bloodstream isolates in North America. International Journal of Antimicrobial Agents. 2002, 20(6), 412-418.

9. **Gamaletsou M.N., Walsh T.J., Zaoutis T., Pagoni M., Kotsopoulou M., Voulgarelis M., Sipsas N. V.** A prospective, cohort, multicentre study of candidaemia in hospitalized adult patients with haematological malignancies. Clinical Microbiology and Infection. 2014, 20(1), 050-057.

10. **Leung A.Y., Chim C.S., Ho P.L., Cheng V.C., Yuen K.Y., Lie A.K., Au W.Y., Liang R., Kwong Y.L.** Candida tropicalis fungaemia in adult patients

with haematological malignancies: clinical features and risk factors. J Hosp Infect. 2002, 50, 316-319.

11. Nucci M., Colombo A. L. Candidemia due to Candida tropicalis: clinical epidemiologic and microbiologic characteristics of 188 episodes occurring in tertiary care hospitals. Diagn Microbiol Infect Dis. 2007 58, 77-82.

12. Yang Y.L., Cheng M.F., Wang C.W., Wang A.H., Cheng W.T., Lo H.J. The distribution of species and susceptibility of amphotericin B and fluconazole of yeast pathogens isolated from sterile sites in Taiwan. Med Mycol. 2009 ; 1-7.

13. Kothari A., Sagar V. Epidemiology of Candida bloodstream infections in a tertiary care institute in India. ndian J Med Microbiol Apr-Jun 2009, 27(2):171-2.

14. Verma A.K., Prazad K.N. Singh M. et al. Candidaemia in patients of a tertiary health care hospital from north India ; Indian J. Med Res. 2003, 117:122-8.

15. Sangaré I., Sirima C., Bamba S., Zida A., Cissé M., Bazié W.W., Sanou S., Dao B., Menan H., Guiguemdé R.T. Prevalence of vulvovaginal candidiasis in pregnancy at three health centers in Burkina Faso. Myc Med N°726. 2017; 7p

16. Jia-Yun Tsai C., San-Loh J. M., Proft T. Galleria mellonella infection models for the study of bacterial diseases and for antimicrobial drug testing, Virulence. 2016, 7:3, 214-229.

17. De-Dong Li, Li Deng, Gan-Hai Hu, & al. Using Galleria mellonella-Candida albicans Infection Model to Evaluate Antifungal Agents. Biol. Pharm. Bull. 2013, 36(9) 1482-1487

18. Barnett J.A., Payne, R.W. and Yarrow D. Yeasts: Characteristics and Identification. Cambridge University Press 3rd edition. 2000.

19. Gary C., Sean D., Kavanagh K. Development of an insect model for the in vivo pathogenicity testing of yeasts. FEMS Immunology and Medical MicrobiologyN°27. 2000, 163-169.

20. James T.Y., Kauff F., & al. Reconstructing the early evolution of Fungi using a six-gene phylogeny. Nature 2006, 443 :818-22.

21. Calderone R.A. and Fonzi W.A. Virulence factors of Candida albicans. Trends Microbiol. 2001, 9(7): 327-35.

22. Silva S., Melyssa M., Henriques M., & al. Candida glabrata,Candida parapsilosis and Candida tropicalis: biology, epidemiology,pathogenicityand antifungal resistance. FEMS Microbiol Rev 36.2012 ; 288-305.

23. Fitzpatrick D.A., Logue M.E., Stajich J.E., Butler G. A fungal phylogeny based on 42 complete genomes derived from supertree and combined gene analysis. 2006; BMC Evol Biol 6 :99.

24. Gow N.A.R., Van der Veerdonk F.L., Brown A.J.P, Netea M.G. Candida albicans morphogenesis and host defence: discriminating invasion from colonization. Nat Rev Micro. 2012 Feb; 10(2):112-22.

25. Norice C.T., Smith F.J., Solis N., Filler S.G., Mitchell A.P. Requirement for Candida albicans Sun41 in biofilm formation and virulence. Eukaryotic Cell. 2007 Nov;6(11):2046-55.

26. Felk A., Kretschmar M., Albrecht A., & al. Candida albicans Hyphal Formation and the Expression of the Efg1-Regulated Proteinases Sap4 to Sap6 Are Required for the Invasion of Parenchymal Organs. Infect Immun. 2002 Jul 1; 70(7):3689-700.

27. Martin R., Wachtler B., Schaller M., Wilson D., Hube B.H. pathogen interactions and virulence-associated genes during Candida albicans oral infections. International Journal of Medical Microbiology. 2011 Jun 1;301(5):417-22.

28. Shapiro R.S., Cowen L.E. Uncovering cellular circuitry controlling temperature-dependent fungal morphogenesis. Virulence 2012; 3:400-4;

PMID:22722238;

29. Mayer F.L., Wilson D., Hube B. Candida albicans pathogenicity mechanisms. virulence. 2013 Feb 15; 4(2):119-28.

30. Brandhorst, T. T., Wuthrich, M., Finkel-Jimenez, B., Warner, T., & Klein, B. S. Exploiting Type 3 Complement Receptor for TNF- Suppression, Immune Evasion, and Progressive Pulmonary Fungal Infection. The Journal of Immunology. 2004 ;173(12), 7444- 7453.

31. Pereira T., de Barros P., Fugisaki L., & al. Recent Advances in the Use of Galleria mellonella Model to Study Immune Responses against Human Pathogens. J Fungi. 2018; 4: 128.

32. Ellis J.D., Graham J.R., Mortensen A. Standard methods for wax moth research. Journal of Apicultural Research. 2013;52, 1-17.

33. Ramarao N., Niclsen-Leroux C., Lereclus D. The Insect Galleria mellonella as a Powerful Infection Model to Investigate Bacterial Pathogenesis. Journal of Visualized Experiments. 2012; 1-7.

34. Jorjao A.L., Oliveira L.D., Scorzoni L., & al. (2017) From moths to caterpillars: Ideal conditions for Galleria mellonella rearing for in vivo microbiological studies. Virulence, 5594, 00-00.

35. Kwadha C.A, Ong'Amo G.O., Ndegwa P.N., Raina S.K., Fombong A.T. The biology and control of the greater wax moth, Galleria mellonella. Insects. 2017; 8, 1-17.

36. Sheehan G., Garvey A., Croke M., Kavanagh K. Innate humoral immune defences in mammals and insects: the same, with differences? Virulence 2018; 9: 1625-39.

37. Kavanagh K., Reeves E. Exploiting the potential of insects for in vivo pathogenicity testing of microbial pathogens. FEMS Microbiol Rev 2004; 28: 101-12.

38. Tsai C.J., Loh J.M., Proft T. Galleria mellonella infection models for the study of bacterial diseases and for antimicrobial drug testing. Virulence. 2016 ; 7: 214-29.

39. Reeves E.P., Messina C.G.M., Doyle S., Kavanagh K. Correlation between gliotoxin production and virulence of Aspergillus fumigatus in Galleria mellonella. Mycopathologia. 2004; 158:73-9.

40. Sudbery PE. Growth of Candida albicans hyphae. Nat Rev Microbiol. 2011 Oct; 9(10):737- 48.

41. Raneri, M., Pinatel, E., Peano, C., Rampioni, G., Leoni, L., Bianconi, I., . Briani, F. Pseudomonas aeruginosa mutants defective in glucose uptake have pleiotropic phenotype and altered virulence in non-mammal infection models. Scientific Reports. 2018 8(1).

42. Imran M, Desmasures N, Coton M, et al. Safety assessment of Gram-negative bacteria associated with traditional French cheeses. Food Microbiol 2019; 79: 1-10.

43. Accoceberry I. and Noel T. "Antifungals cellular targets and mechanisms of resistance." Therapie 2006: 61(3): 195-9.

44. Mesa-Arango A.C, Trevijano-Contador N., Roman E., Sanchez-Fresneda R., Casas C., Herrero E., Arguelles J.C., Pla J., Cuenca-Estrella M., Zaragoza O. The production of reactive oxygen species is a universal action mechanism of amphotericin B against pathogenic yeasts and contributes to the fungicidal effect of this drug. Antimicrob Agents Chemother. 2014; 58: 6627-6638.

45. Young L.Y., Hull C.M., Heitman J. Disruption of ergosterol biosynthesis confers resistance to amphotericin B in Candida lusitaniae. Antimicrob Agents Chemother. 2003; 47:2717-2724.

46. Perini Hugo F., Alane T.P., Moralez R.S.C & al. Phenotypic switching in Candida tropicalis alters host-pathogen interactions in a Galleria mellonella infection model. In Scientific Reports, published online : 29 August 2019.

47. Koenig H. Guide to Medical Mycology. Collection Ellipses; 1995.

48. Shokal U, Eleftherianos I. Evolution and function of thioester-containing proteins and the complement system in the innate immune response. Front. Immunol. 2017; 8.

49. Mesa-Arango A.C., Forastiero A., Nez L. B., Cuenca-Estrella M., Mellado E. & Oscar Zaragoza O. The non-mammalian host Galleria mellonella can be used to study the virulence of the fungal pathogen Candida tropicalis and the efficacy of antifungal drugs during infection by this pathogenic yeast. Mycology Reference Laboratory, National Centre for Microbiology. Colombia; in Medical Mycology July 2013, 51, 461-472.

50. Kafando C. In vivo study of the virulence of clinical strains of Candida albicans resistant to fluconazole or amphotericin B according to the Galleria mellonella model. Doctoral thesis in Pharmacy, Ouagadougou (Burkina Faso); May 2020, N°199, 146p.

51. Brennan M., Thomas D.Y., Whiteway M., Kavanagh. Correlation between virulence of Candida albicans mutants in mice and G. mellonella larvae. FEMS Immunology and edical Microbiology 34. 2002 ; 135-157

APPENDICES

Appendix 1: Protocol for growing soil samples

Stage 1: Culture in YPD/agar medium+ Ampicillin+ Chloramphenicol+ Kanamycin

- Incubate each sample in 10 mL YPD + 100 Ug/UL ampicillin (ampicillin preparation, 100mg in 1mL or 500UL distilled water and 500UL ethanol= 100mg/mL)+ 50 Ug/L kanamycin and 50 Ug/L chloramphenicol.
- The samples are incubated with the control

- Incubation : 30°C

- Observation of growth. After 24 hours, turbidity of the medium compared to the control (no turbidity).
- Leave to grow for a further 24 to 48 hours

Step 2: Reproducing de novo in YPD/agar

Step 3: Transplant into sabouraud gelose (chromogenic medium)

Step 4: Identification

- Maldi-tof

- PCR (Polymerisation chain reaction)

Appendix 2: Technique for filling the Malassez cell with strain suspensions

After preparing the suspension of the C. tropicalis strain, we homogenised the suspension and sampled it using a Pasteur pipette, filling the counting chamber by capillary action. Filling was done in a single operation, without air and without overflowing the liquid into the channels.

✓ **Counting C. tropicalis cells on the Mallassez cell**

We first observed at the x10 objective to locate the position and check the homogeneity of the distribution of the cells to be counted, then we observed at the x40 objective to count the C. tropicalis. After counting, we calculated the fungal load of the C. tropicalis strain suspension using the formula below:

$$\mathbf{N = (n / v).f}$$

With : - **n**= number of C. tropicalis counted,

- **v**= volume of yeast suspension,

-**f**= dilution factor,

N= number of C. tropicalis per microlitre (the volume corresponding to the total grid of the Malassez cell is equal to 1µL).

✓ **Dilution method for cell suspensions**

After determining the fungal load of the suspension of the telluric strain, we carried out a series of dilutions at varying concentrations of 5.10^5; 1.10^6 and 5.10^6 CFU using the formula below:

$$C1.V1 = C2.V2$$

With :- **C1**= the initial concentration of the cell suspension

V1= the initial volume of the cell suspension

-C2=the final concentration of the cell suspension

V2= final volume of cell suspension.

Numerical application: S1R1Y1 strain

- We counted 25,000 C. tropicalis
- The solution was diluted 1:10^{e}
- The counting volume is equal to 1µL
- The quantity of C. tropicalis suspension to be inoculated into each larva is 20µL.

Numerical application: **N**= (25,000 C. tropicalis/1µL) x 10 x 20µL= **5.10^{6}UFC/20µL**

The concentration of the S1R1Y1 stock suspension= 5.106CFU/20µL

Appendix 3: Photos

Oviposition of G. mellonella females: several eggs laid on white paper.

A: Ingredients needed to prepare food for G. mellonella

B: repair of food for

Dry ingredients (200 grams of wheat flour and 300 grams of oats) were mixed with 10 grams of honey.

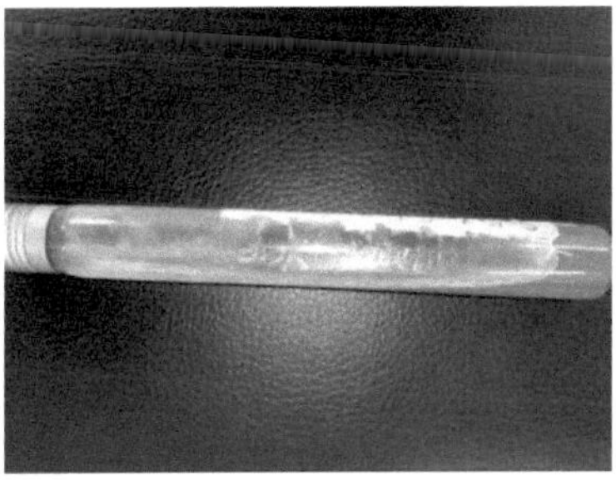

Photo: C. tropicalis strain S1R1Y1 on SC medium

SUMMARY

Title: *In vivo study of the virulence of a telluric strain of Candida tropicalis using the model Galleria mellonella*

Introduction :

The incidence of fungal infections in humans has increased over recent decades, ranging from superficial to systemic cases. Candida tropicalis is one of the most dreaded non-albicans Candida species associated with deep and systemic candidiasis, accounting for 3 to 66% of blood isolates worldwide. In Burkina Faso, studies of fungal virulence using the invertebrate model are still at a preliminary stage. The aim of our study is to assess the in vivo virulence of a telluric strain of C. tropicalis resistant to amphotericin B using the G. mellonella model.

Materials and methods

This was an experimental prospective study. To estimate virulence, batches of 10 healthy larvae weighing between 280 and 320mg were inoculated with a telluric strain of Candida tropicalis identified by MALDI-TOF. The larvae were inoculated with 20µl of cell suspension at increasing concentrations of 5.10^5, 1.10^6, and 5.10^6 CFU. A control group injected with 20µL of sterile phosphate buffered saline (PBS) was included. After incubation at 37°C, melanisation and larval death were observed every 24 hours. The experiment was terminated after 12 days; a strain was considered virulent if it caused 100% larval mortality over a maximum of 96 hours. The results were analysed using the Kaplan-Meier method on XLSTAT 2020.1.2; the difference was considered significant for $p<0.05$.

Results

The strain studied was resistant to amphotericin B but sensitive to six other antifungal agents.One hundred per cent of the larvae melanised 30 minutes after inoculation at concentrations of 1.10^{6}, and 5.10^{6} CFU/larva. One hour later, all the larvae underwent melanisation except for the single PBS-injected group in which no larvae underwent melanisation. No mortality was observed in the PBS group; however, 100% mortality was observed after 48 hours ($p<0.001$) in the two groups that received a fungal load of 1.10^{6} and 5.10^{6} CFU/larva. After 12 days, 20% of larvae in the 5.10$^{(5)}$ group survived ($p=0.005$). The C. tropicalis strain is therefore not virulent at a concentration of 5.10^{5} CFU; we estimate that a minimum concentration of 1 10^{6} CFU of the strain is required to be virulent.

Conclusion

Candida tropicalis is one of the causes of serious infections. Amphotericin B, the antifungal agent of choice for the treatment of deep-rooted candidiasis, is proving to be ineffective against our strain. The G. mellonella model, which is easy to use, could serve as a Gold Standard for experiments on the virulence of fungal microorganisms.

Key words: *virulence, in vivo, strain, telluric, amphotericin B, Candida tropicalis, Galleria mellonella, MALDI-TOF, haemocytes, CFU, Burkina Faso*

Author : *Hermann Jonas TAPSOBA ; Tel: (00226) 70667885/64229901*

_Email :_*ghermannjonast@gmail.com*

MIX
Papier aus verantwortungsvollen Quellen
Paper from responsible sources
FSC® C105338

Printed by Books on Demand GmbH, Norderstedt / Germany